GLACIER ICE

Austin Post *and* Edward R. LaChapelle

UNIVERSITY OF TORONTO PRESS

Frontispiece: Mount Hayes, Alaska Range

GLACIER ICE

Preface

In 1955 the first organized aerial photography of glaciers in western North America was started under the auspices of the University of Washington by Richard C. Hubley. His photographic reconnaissance of the North Cascades brought to public attention the extent of glacierization in this area and identified numerous glacier advances generated by a moderate trend toward a cooler and wetter climate. Hubley repeated this survey each year until his untimely death in northern Alaska in October, 1957, during the International Geophysical Year. LaChapelle continued the Cascade Range photo flights through 1960 in order to maintain the record of glacier fluctuations there.

Beginning in 1960, Post expanded the annual photo-survey of glaciers along the north Pacific Coast of North America and into the interior ranges of Alaska. This work was supported by National Science Foundation grants to the University of Washington. In 1964 Post joined the Water Resources Division of the United States Geological Survey. Since that time the photographic reconnaissance of glaciers in western North America has been continued by the survey and has expanded to include other mountain ranges. The extent and character of glacierization has now been well documented, the patterns of glacier change are beginning to emerge, and the extent of phenomena like glacier surges has been established.

These aerial reconnaissance techniques have now become firmly established as a tool of glaciology. A by-product is the accumulation of a very large file of aerial photographs of glaciers, a file that is augmented by hundreds of new negatives each year. As we have catalogued and studied these photographs, continuing our scientific research on glacier extent and behavior in North America, we have encountered outstanding examples of many well-known (and some not so well-known) glacier phenomena. This suggested at an early stage that material was at hand for an illustrated text on glacier features.

The majority of photographs in this book are taken from Post's aerial photograph files. As some features could best be illustrated by a ground-based camera, we have included numerous other photographs taken in the course of our glacier research projects. Inquiries among our colleagues led to additional illustrations of certain features missing from our own files. The final compilation that appears here is thus gathered from several sources, with Post's aerial photographs as the foundation.

There is a hidden contribution to most of these photographs which is not obvious to the armchair reader, but which we wish to emphasize and acknowledge. Aerial photography of glaciers requires not only a properly equipped photographer—it is a team effort, and the other half of the team is the aircraft pilot. Weather conditions over the mountains of the North Pacific Coast are often poor, and good picture weather is rare. The glacier areas are remote, navigation aids are scarce, and the terrain is as inhospitable to lost fliers as any in the world. Turbulent air and sudden downdrafts over the mountains can treat a light aircraft to unexpected and awesome roller-coaster rides. A high order of piloting skill and a thorough knowledge of the mountains and weather are essential to a safe return. In addition, the pilot must understand and even anticipate the desires of the photographer so that he can position the aircraft for the best pictures without wasting time and fuel. Many glacier photographs taken on the ground also depend on a skillful pilot, for much modern glaciology relies heavily for logistic support on ski-wheel aircraft which can land and take off under adverse weather and snow conditions from tilted and formidably short sections of glaciers.

For both our aerial photographs and our air-supported glacier research projects, the pilot was the same: William R. Fairchild of Port Angeles, Washington. Without his remarkable abilities most of the photographs in this book would never have been taken.

On February 5, 1969, Bill Fairchild was killed in a tragic crash while taking off from the Port Angeles airport.

To the memory of Richard C. Hubley, brilliant glaciologist and originator of aerial glacier studies, and of William R. Fairchild, the pilot whose skill made these projects possible, this book is sincerely dedicated.

Austin Post
Edward R. LaChapelle

Contents

Illustrations

GLACIER ICE

1. *Crystals of lake ice (candle ice)*

2. *Snow crystal*

Glacier Formation and Mass Balance

When water freezes in bulk, the shape of the ice crystals formed depends on the rate of freezing and the manner in which the water is exposed to low temperatures. On the surface of the earth, this process commonly takes place on the open surface of oceans, lakes, and ponds exposed to cold air. In fresh water the characteristic *candle ice* crystals of lake ice are formed. These columnar crystals, which form at right angles to the water surface and often reach several centimeters in length, are not readily discernible in the solid sheet of ice covering a lake, but they become visible under favorable conditions when sunlight melts the crystal boundaries and disaggregates the ice into individual crystals. Such typical candle ice is seen in Figure 1.

Water vapor in the atmosphere also freezes, first forming tiny ice crystals which under various conditions grow to become snow crystals of almost infinite variety. A *glacier* is formed on land in areas where more of these snow crystals accumulate annually than melt. Over a period of many years these accumulations become deep enough to form glacier ice, which by a number of processes flows because of its own weight. A glacier is most simply defined as a mass of perennial ice on land that shows evidence of glacier flow.

The crystals of glacier ice thus are not "frozen water" in the sense of lake ice. That is, they did not gain their present form by the freezing of liquid water. Instead, these crystals are the product of extensive metamorphism. Snow crystals like those shown in Figure 2 have a structure entirely different from that ultimately exposed at a glacier surface. The conversion to glacier ice may be aided initially by melting and refreezing in the surface snow layers, but it is the long, slow process of metamorphism, water molecule exchange from one ice particle to another, that eventually over years, decades, and centuries converts the original tiny snow particles into large glacier crystals. The ultimate size of individual crystals is influenced by glacier flow as well as age; under favorable circumstances they may reach 30 centimeters or more in diameter. Melt by the sun also disaggregates the surface of glacier ice, but the crystal shape revealed, such as shown in Figure 3, is entirely different from that of lake ice (Fig. 1). Each winter a mountain glacier is blanketed with fresh accumulations of snow that may reach depths of as much as 6 to 10 meters in a maritime climate. The thickness of this accumulation usually increases with altitude up to the level of maximum precipitation, which generally occurs at altitudes of 1,000 to 3,000 meters along the north Pacific Coast. In the summer much of this new snow is removed by *ablation* (melting), the rate of melt usually decreasing with altitude. As a result of these altitude variations, the winter snow vanishes first from the glacier terminus, and then from successively higher altitudes on the glacier. The boundary between the winter snow accumulation and the bare ice of the lower glacier exposed by summer melt is called the transient snowline, or simply the *snowline* (see Figs. 5 and 35). This should not be confused with the *regional snowline*, which is defined as the altitude at which annual accumulation balances ablation on a ground surface. Accumulated snow and its lower limit, the snowline, persist on a glacier long after snow has vanished from the surrounding rocks, for unbroken snow surfaces are much less sensitive to melt by solar radiation than are those interrupted by rocks and trees.

3. *Crystals of glacier ice*

4. *Wind-controlled snow accumulation, unnamed glacier, Kenai Mountains*

Snow that survives a year or more of ablation on a glacier is called *firn*. The contact between bare ice and firn is called the *firn edge*. At the end of the summer the snowline reaches some maximum altitude on the glacier. This is the *firn line*. Above it is the *accumulation area* where more snow is deposited than melts year after year and is slowly compressed into ice. Below the firn line is the *ablation area* where all annual snow melts most years and ice is also ablated away. Under the influence of gravity, ice flows from the accumulation to the ablation region. Under ideal steady state conditions the snow accumulation, ice flow, and snow and ice melt all balance so that no change in mass occurs. As such conditions are rare in nature, glaciers are constantly adjusting to changes in snow accumulation and ablation by increasing or decreasing their flow velocity and by advance or retreat of the *terminus*, or lowest point of the glacier.

5. *Snowline, South Cascade Glacier, North Cascades, 1957*

Snow accumulation on a glacier is irregular, being much thicker one place than another even at the same altitude. The relation of local topography to prevailing winter storm winds determines which part of the glacier will receive the deepest snow deposition. Figure 4 shows an extreme example of terrain effects on wind-drifted snow. Summer ablation also takes place unevenly, for it is affected by the relation of local topography to sun and warm winds. The variations of accumulation and melt do not coincide. These differences are expressed by the snowline and firn line mentioned above. The example illustrated in Figure 5 is a simple case; often the firn edge is highly irregular when the ablation surface intersects uneven accumulation horizons. When exceptionally heavy summer melt occurs, more than one season's accumulation may be intersected. In the North Cascades severe melting occurred in 1958, and the South Cascade Glacier presented a strikingly different firn edge picture in Figure 6a and b. Snow layers from six different years of accumulation were exposed when these photographs were taken.

a

b

6a,b. Two views of firn edges, South Cascade Glacier, North Cascades, 1958

A similar feature in an entirely different climate appears in Figure 7. These small, stagnant glaciers are found in Axel Heiberg Island in the Canadian Arctic Archipelago. Annual accumulation here is very small compared with the Cascade Range of Washington. At such a high latitude (almost 80° N), ablation is also much lighter. But a similar pattern of firn edge has developed, probably for the same reason as on the South Cascade Glacier. The surrounding area exhibits *soil polygons,* a type of patterned ground common to permafrost areas at high latitudes.

7. Small ice patches on Axel Heiberg Island, Northwest Territories

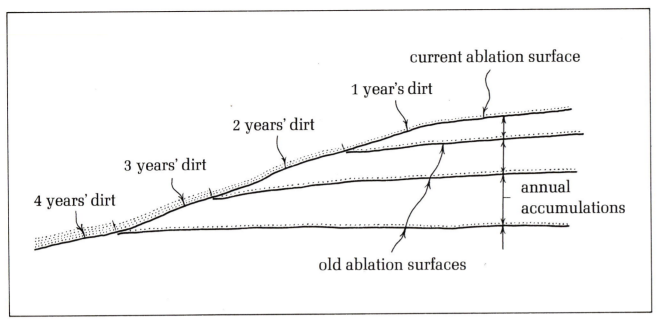

current ablation surface

1 year's dirt

2 years' dirt

3 years' dirt

4 years' dirt

annual accumulations

old ablation surfaces

8. *The formation of ablation surfaces*

The identification of each annual firn layer in such a picture is based on the dust and dirt that accumulate on the glacier each year. Such identification is not possible in very clean, dirt-free snow, such as that found on the dust-free central ice sheets of Greenland and Antarctica. But in alpine regions the wind picks up dust from both near and distant bare ground and deposits some of this along with snow on the glacier. Such dust is thinly distributed throughout the winter snowpack and is not visible in winter or early spring when the snow presents a clean white surface. As spring and summer melt proceed, this dirt and dust accumulate on the surface. An *ablation surface* thus develops, with its characteristic surface dirt layer, which becomes darker as summer progresses. Such an ablation-formed dirt horizon divides one year's firn accumulation from the next. When melt is strong enough to remove entirely the annual snow, its dirt layer is added to the next layer underneath. If ablation continues, this process can result in the dirt from several years' accumulation collecting together, each addition giving the exposed surface a successively darker color as shown in Figure 8. The many shades of surface "dirtiness" (seen in Fig. 6a and b) have been formed in just this fashion, and the pattern of their boundaries reveals clearly the characteristic distribution of snow accumulation on this glacier.

In many ways the formation of glacier ice is analogous to the formation of metamorphic rock structures in the earth's crust. (From the geological viewpoint, in fact, ice is a rock—simply one with a much lower melting point than most.) Instead of lime, silt, and other rock materials being deposited by sedimentation on a lake or ocean bottom, a layer of snow is deposited by sedimentation from the atmosphere. Successive layers compress those beneath, and metamorphism, influenced by percolating meltwater and prevailing temperatures, gradually converts the snow crystals to new forms, just as it does rock particles. The stratigraphy, or layering, in firn and ice may remain visible for many years wherever a section is exposed by fracturing or erosion. Glacier flow first distorts this layering and then completely obscures it, although visible layers will persist a long time if the glacier motion is simple and unbroken. The layering seen in Figure 9 has been tilted as it was carried down to the zone of melt, but is still clearly recognizable. This can happen only if the course of the glacier is free of ice cascades or other irregularities and little shearing takes place.

9. *Accumulation layering in glacier ice, Chakri Glacier, Garwhal Himalayas*

10. Accumulation layering in firn

Accumulation layers are also visible in Figures 10 and 11. Such layers, their form distorted by flow and by ablation, can also be recognized in Figure 71. Figure 11 exhibits a complex record of weather influence. Most of the ice in this exposed face was deposited as shallow layers which have been deformed downward by flow at the center to form gentle curves. At the point indicated by the arrow, an abrupt transition in the layering occurs. Identical transitions are found in sedimentary rocks, where they are known as *unconformities*. In the case of both rock and ice, they represent an interruption in the regular, annual sequence of accumulation. Exposure for a time to erosion (melting in the case of ice) results in a surface which may not necessarily conform to the deposition surface. Subsequently, when accumulation resumes, layers are deposited whose surfaces conform to the intervening erosion surface. In the case of the Blue Glacier shown here, recent climatic changes stand clearly revealed. The erosion surface represents a period of intensive glacier melting known to have occurred in the Pacific Northwest during the 1930's and 1940's. Subsequent deposition of the nearly horizontal layering above the arrow occurred during the recent period of glacier rejuvenation beginning in 1948. Ice below the unconformity must therefore have been laid down as snowfalls before the 1930's. Presumably the layering displayed here is annual, hence the ice at the foot of this face must date from the turn of the century or earlier.

Not all glaciers exhibit a primary layering built up by annual snow deposition from the atmosphere. Some, for instance, consist almost entirely of ice formed by refrozen meltwater that percolated down from surface snow melt. A rather critical combination of annual snow accumulation and seasonal mean temperature variations is required for this. Others are nourished mostly by snow avalanches sliding from surrounding steep peaks, a form of secondary deposition which accumulates a jumbled mass of snow with much less distinct layers. Figure 12 illustrates terrain typical of avalanche-fed glaciers.

Glacier ice, like rock, preserves in its layering the history of weather influences that went into its making. Sometimes unusual events will leave a special mark in the stratigraphic profile; such an example is indicated by arrows in Figure 13. This small glacier is situated on the Alaska Peninsula, where active volcanoes abound. In the recent past a volcanic eruption blanketed the land with a layer of ash, and this ash layer was then sandwiched between subsequent snowfalls on the glacier. Slow ice movement has now carried it down past the firn limit, where the erosion surface of the lower glacier has intersected the accumulation layering at an angle and revealed the edge of the ash layer as a dark band, here still partly obscured by the annual snowline. No scientist has yet visited this nameless glacier to determine the age of the ash layer, but, since it is not deeply buried, it must date from recent years. Presumably it is derived from nearby Chiginagak Volcano (Fig. 100).

A similar ash layer of earlier origin (probably the Mount Katmai eruption of 1912) appears in Figure 14, which shows an outlet glacier draining the Harding Ice Field. This layer was incorporated into the accumulation zone long ago and has been transported by glacier flow almost to the terminus, where extensive ablation has diffused the ash over the underlying ice. The upper ash boundary is still sharp where it has been protected by the layers of ice on top of it.

14. *Ash layer preserved in ice stratigraphy, unnamed glacier in Harding Ice Field, Kenai Mountains*

15. *Accumulation survey, Blue Glacier, Olympic Mountains*

The annual mass balance of a glacier is the difference between the amount of ice it gains each year (usually as snowfall) and the amount it loses (usually as melt). In order for a glacier to grow, it must have a sustained positive mass balance (net mass gain each year). If the mass balance is persistently negative, the glacier will shrink. The relations between a glacier's mass balance and its terminal advance or retreat are complex. Considerable time lag may exist before the one has an effect on the other. The state of health of a glacier (the character of its mass balance) is not always obvious from the behavior of its terminus. The position of the snowline is a better indicator, but it does not by any means tell all the story. An accurate assessment of a glacier mass balance can be gained only by actually measuring the ablation and accumulation. Part of the accumulation-measuring process, determining the water content of snow, is shown in Figure 15.

When does a snowfield reach a sufficient size to become a glacier? Or, conversely, when does a re-treating glacier cease to be one? To qualify as a glacier, a body of ice ought to have at least some true glacier ice, whose pore volume is no longer communicating like that of snow. It ought to exhibit glacier flow, but this immediately raises a problem, for even a winter snow cover is subject to rapid deformation under the influence of gravity. Ice motion exhibited by at least one *crevasse* somewhere on the glacier is a requirement demanded by many glaciologists. The geologist, on the other hand, may point to geomorphological features, the effects of ice on the landscape such as *moraines* or *striae* (see Figs. 45 and 105), as the positive indications of a glacier flow and hence a real glacier.

An example of a marginal glacier appears in Figure 16. The snow patch might be a perennial feature, or just one that happened to survive *this* particular ablation season. Old snow or ice could be hidden underneath the mantle of rock debris. Insignificant as the snow patch in Figure 16 may be, there still is no disregarding that large, obvious *terminal moraine* beneath it. In fact it is that moraine which raises the problem of identification in the first place; without it this tiny ice mass would be considered to be no more than a lingering snow patch.

16. Unnamed glacier, Wasatch Mountains

17. Bergschrunds, Mount Ratz, British Columbia, Coast Mountains

Major Surface Features of Ice

At the head of most alpine glaciers (as distinguished from ice sheets) there is a zone where snow accumulation rapidly diminishes with altitude, often where the glacier approaches a wind-swept ridge or a rock wall too steep to support an ice mantle. Within this zone active flow is not possible, for too little snow accumulates each year. Below this zone the normal process of mass accumulation and flow sustains an appreciable glacier motion. A persistent feature of most active glaciers is the *bergschrund,* a large crevasse which separates the flowing ice from the nearly stagnant ice mass above. Often it runs clear across the head of the glacier, where it constitutes a late-summer obstacle to mountaineers when snow bridges have melted away.

The bergschrund should be distinguished from another glacier feature whose name also originated in the Alps. This is a *randkluft,* the space or crack between the edge of the glacier and the adjacent rock wall. It is bounded on one side by ice and on the other by rock, while the bergschrund divides two ice masses. The randkluft is developed predominantly by melt adjacent to the dark rock, rather than by flow.

Bergschrunds are visible in many illustrations in this book. Several are especially clear in Figure 17, where they are indicated by arrows. Above the right-hand arrow the stagnant ice mantle has thickened until it has begun to flow and form new cracks above the present bergschrund. Eventually these new cracks will become the bergschrund, while the old one below closes and is carried down the glacier.

Figure 18 provides another illustration of glacier accumulation areas. The basin in the middle of this picture shows by its heavily crevassed and broken surface that it is in rapid motion and discharging a large amount of ice each year by vigorous flow. The quantity of ice any glacier must transport by flow each year from a higher to lower altitude depends on the rate at which mass is added by snowfall above and removed by melt below. The *activity index* of a glacier is the rate of increase of net mass balance with altitude. The larger the activity index, the greater is the annual volume of ice that must be transported across the firn line by glacier flow. Active glaciers with vigorous flow are most common in maritime climates where snow lies deep in the high mountains and summer melt is strong at lower elevations. Glaciers in a continental climate usually have a much lower activity index.

Seen from the ground, the broken surface of an accumulation area such as that in Figure 18 is a chaotic jumble of crevasses and *seracs,* or isolated ice blocks, which seriously impede travel by mountain climbers. Passage through such a zone may be possible early in the melt season while deep snow still blankets the obstacles (Fig. 10), but as summer melt erodes the blocks and uncovers the crevasses, travel becomes increasingly difficult (Fig. 19).

To many people, a glacier means ice—bare, blue ice. This expected aspect of a glacier is found only below the firn line in the ablation zone. Here the summer melt removes entirely, each year, both the winter snow and some of the ice that has been formed by compaction in the accumulation zone and carried down by glacier flow. In extensive high mountain areas where annual snow accumulation far exceeds melt, the glaciers from many accumulation areas covering hundreds of square kilometers may coalesce to form a single great river of ice as their common ablation zone.

18. *Accumulation zone crevasses, Bishop Glacier, British Columbia, Coast Mountain*

19. Seracs in firn, Blue Glacier, Olympic Mountains

20. *Bering Glacier, south central Alaska*

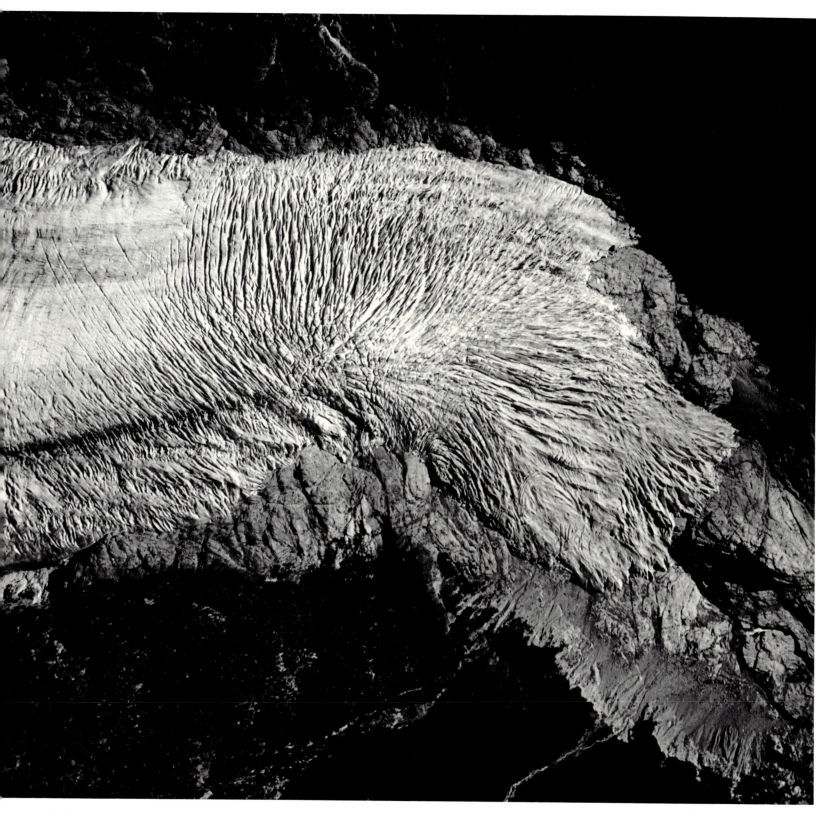

The Bering Glacier in Alaska (Fig. 20) is an outstanding example of such an ice river. The part shown in the photograph was taken where the Bering Glacier is 11 to 16 kilometers wide. Covering an area of over 6,000 square kilometers, this glacier system is the world's largest outside of Greenland and Antarctica.

Because the ablation zone surface is not buried by snow accumulation throughout each year, it can display clearly the crevasse patterns and other features that persist from year to year. The lower reach of the Sumquolt Glacier shown in Figure 21, which has recently experienced retreat, shows a strong fracture pattern of crevasses generated by flow of the ice around a bend in its channel (upper center). To the left of this pattern is a fainter one with fractures at a different angle, formed by a flow feature farther upstream. Where these two patterns intersect (just left of center), a field of seracs is formed. Seen from the ground such ice features are similar to the firn blocks of the accumulation zone, but they consist of solid ice which, as shown in Figure 22, is often eroded and rounded by melt.

Characteristic crevasse patterns are produced by shearing in the glacier ice induced by friction of the valley walls between which the glacier flows. The patterns are different in the accumulation and ablation zones. In the accumulation zone, the glacier flow velocity increases downstream because the amount of accumulated mass that must be transported also increases downstream. Such a phenomenon is called *extending flow*. Combined with side friction, it produces crevasse patterns like the one shown in Figure 79. The reverse situation is found in the ablation zone. Here the amount of mass that must be transported decreases downglacier and so does the flow velocity. This is called *compressing flow*, which combines with side friction to produce *chevron crevasses* like those seen along the margins of the Sumquolt Glacier in Figure 21.

21. *Crevasse patterns, Sumquolt Glacier, British Columbia, Coast Mountains*

The Sumquolt Glacier is relatively slow-moving, with an ablation zone exposed to substantial melt each summer. These factors combine to produce a blurred pattern of seracs and crevasses which does not exhibit all the jagged angularity of freshly fractured ice. On the other hand, in Figure 23 the Gakona Glacier displays a fresh and unobscured pattern of fractured glacier ice. Here the complex strains in the rapidly flowing ice have shattered the glacier surface into orderly arrays of seracs. These seracs have suffered very little melt and have maintained their original form and orientation for some distance downglacier from the point where the ice surface was broken apart. A glacier surface like this is almost impossible to cross on foot.

An *icefall* demonstrates very conspicuous evidence of active glacier flow, and its broken, chaotic surface may be impossible for the most expert mountaineer to traverse. Flow velocity in an icefall may reach a thousand or more meters per year, much faster than in other parts of most glaciers. In spite of the name, this feature is still not an ice "fall" in the sense of an avalanche of ice from the terminal of a *hanging glacier* which cascades down a mountainside in seconds or minutes. The flow of the underlying ice is accelerated by the steep gradient and constricted channel, while the surface layer of the glacier, unable to accommodate to this change, is broken into crevasses and seracs which are pulled apart by the extending flow at the top. Many of the individual seracs and ice blocks do collapse and tumble down, one by one. At the foot of the icefall the motion decelerates, and compressing flow rapidly closes the fissures and restores a relatively smooth ice surface.

22. *Broken ice and seracs, ablation zone, Lemon Creek Glacier, southeast Alaska, Coast Mountains*

23. *Crevasse patterns resulting from very rapid surge, Gakona Glacier, Alaska Range*

24. *Icefalls, Mount Ratz, British Columbia, Coast Mountains*

25. Icefall, Margerie Glacier, Fairweather Range, Glacier Bay

26. Khumbu Icefall, Mount Everest

Icefalls frequently separate plateaus or basins, which serve as accumulation zones, from the valley ablation areas of mountain glaciers. The series of steep ice cascades in Figure 24 is an extreme example. Another spectacular example is the icefall of the Margerie Glacier in Figure 25, a steep zone in a valley glacier where it descends from the generating firn fields located on a large, upland plateau.

From the ground, icefalls appear just as rough as they do from the air. Normally they are avoided as a route of mountain ascent, both because they are difficult to travel and because of danger from falling seracs. Occasionally no alternate route is available and climbers are forced to travel paths like the famed Khumbu Icefall on Mount Everest (Fig. 26).

a

27a,b,c. Time-lapse sequence of icefall, Blue Glacier, Olympic Mountains

Glacier Flow

In order to produce the characteristics of glaciers—mass transfer, crevasses, erosion, moraine formation—the ice must flow. Taken from the simplest viewpoint, glacier ice is a substance that flows down the mountainside much as a stream of water does, but at a far slower rate. The plastic qualities of ice can be measured in the laboratory, but these are difficult to apply in a simple fashion to calculate glacier behavior. The large-scale motion of a temperate glacier (at the melting point throughout) depends on a complex combination of slip along the bed, fracturing, and ordinary plastic deformation.

Glacier motion also appears to depend to some extent on the amount of meltwater present within and under the glacier. Some of these complexities of motion are suggested in Figure 27a, b, and c. The changes over a period of 2½ months are due not only to flow but to collapse of seracs and ablation. Analysis of the complete series of time lapse photographs from which these are selected shows that individual ice blocks flow sideways as well as down the icefall. The motion is so erratic that some points actually descend in a zigzag fashion.

b

c

28. *Glacier sole exposed on icebergs, South Sawyer Glacier, southeast Alaska, Coast Mountains*

At the bottom of a glacier there is a layer of dirty, debris-laden ice called the *glacier sole*. Rocks of all sizes imbedded in this layer provide the abrasive tools with which the glacier scours and erodes its bed. Pure ice alone would have far less erosive effect on solid and massive bedrock. Deformation of ice in which the rocks are embedded allows it to conform to the shape of the glacier bed and develop a grooved pattern on its under surface.

Access to the bottom of a moving glacier is difficult, and the available photographs of a glacier sole are very few indeed. An aerial photograph is hardly the place where one would expect to find pictured the underside of a glacier, but a fortunate chance occurrence has provided in Figure 28 just such an illustration. The South Sawyer Glacier at the head of Tracy Arm, a long fjord in southeastern Alaska, had just discharged an iceberg at the time this photograph was taken, and several pieces from the bottom of the glacier (indicated by arrows) had floated to the surface and overturned to expose the sole. The photographer was not aware of his luck when he made the exposure; only later did more leisurely examination of the finished print reveal this unusual feature.

The manner in which the grooved undersurface of the glacier sole is generated is seen in Figure 29, another photograph obtained by a chance stroke of luck when glaciologist F. Mueller discovered a cavern in the Pumori South Glacier (near Mount Everest) which gave direct access to the glacier bed. The sliding ice is imprinted with large and small grooves by each bump and irregularity in the bedrock boss at the picture center. A glacier like this slides over the ground by two processes. Plastic deformation of the ice allows it to bend and slide around large obstacles, but it is too stiff to bend around small ones a few centimeters high. Instead, it slides over these by temporarily reverting to liquid water, a process known as *regelation*. The ice melts as a result of pressure on the uphill side of small obstacles, flows around them as water, and refreezes at the downhill side, accompanied by a countercurrent of heat which sustains the process. If a space exists on the downhill side, the refrozen water sometimes forms a feathery deposit of *regelation spicules*. These spicules festoon the upper part of the rock in the center of Figure 29.

29. Glacier sole and grooves, Pumori South Glacier, Garwhal Himalayas

At another part of this same cavern the complex motion of the glacier sole occurs both as fracturing and as plastic deformation around a boulder. Figure 30 shows numerous rocks embedded in the moving ice; these provide the glacier with its tools of erosion.

30. Glacier sole with deformation and englacial debris, Pumori South Glacier, Garwhal Himalayas

31. Glacier sole, folding, and deformation, Pumori South Glacier, Garwhal Himalayas

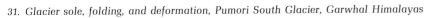

The layer of debris-laden, contorted ice that makes up a glacier sole is seldom as clearly exposed to the light of day as it is along the edge of the Pumori Glacier as shown in Figure 31. The origin of the fold at right is obscure, but this feature offers unmistakable evidence that the process of a glacier sliding over its bed is far from a simple one. The sharp demarcation of the dark, debris-laden sole from the clean glacier ice above is normal.

Glacier Fluctuations

Glaciers are not a fixed feature of mountain landscapes. They wax and wane with changes of climate; even in a stable climate their length may oscillate. For the past century, most mountain glaciers of the world have receded, although local interruptions of the general retreat have appeared from time to time. In a retreating glacier lower reaches are stagnant or flow too sluggishly to replace the ice melted away each summer. The ice shown in Figure 32 slopes gently at the terminus, its surface broken by few crevasses. If the retreat has proceeded far enough, the valley tongue disappears entirely. Figure 33 shows such a decaying glacier, its icefall only a small remnant that reveals a band of bedrock cliffs over which it formerly flowed with enough vigor to fill the valley floor with ice.

32. Retreating glacier tongue, Honeycomb Glacier, North Cascade Range

33. *Desiccated icefall, Coronet Peak, Canadian Rocky Mountains*

The powerful erosive force of an advancing glacier scours soil and bedrock free of vegetation; when the ice retreats, an abrupt transition from bare, mineral rock to mature plant communities is left behind. In Figure 34 this "high ice" mark, called the *trimline,* is especially clear along the margins of Franklin Glacier and its tributaries.

The eroded ground and mantle of debris exposed in the former bed of a retreating glacier is a desolate landscape, at first broken only by *tarns*, small lakes occupying hollows scooped out of the terrain by the glacier flow. At lower altitudes, plants soon appear as the forest begins to re-establish itself on the barren ground, gradually covering the scar and obscuring the trimline (Fig. 35).

34. *Trimlines, Franklin Glacier, British Columbia, Coast Mountains*

35. *Trimlines, tarn, and moraines, Pattullo Glacier, British Columbia, Coast Mountains*

36. *Advancing glacier margin, Taku Glacier, southeast Alaska, Coast Mountains*

Sustained glacier recession in most parts of the world today has accustomed mountaineers and glaciologists to the zone of bare, eroded earth between vegetated ground surfaces and glacier ice (Figs. 34 and 35). These denuded zones are covered with an unstable rock and dirt mantle which is often unpleasant to traverse. It comes as a refreshing change to a footsore mountaineer when he can step directly from clean, blue ice into a mature forest. Such an invasion of timber by advancing ice is rare today, though it must have been common two or three hundred years ago. Many cases of glacier advance are currently being recorded, but most of these have not yet recovered ground lost in recessions of the recent past, so the ice is still surrounded by barren moraines. The Taku Glacier is an outstanding exception. It has advanced steadily for the past fifty years, and today continues the destruction of forests along its margins as shown in Figure 36.

The steep, bulging terminus and a heavily crevassed tongue are both marks of an advancing glacier. The contrast with a receding one is especially clear in Figure 37, where two glaciers of similar shape and size, lying side by side, are in opposite states of behavior.

37. Retreating and advancing glaciers, unnamed glaciers near Alaska-Yukon Border, St. Elias Mountains

The glaciers in Figure 38, in Yukon Territory, have gained the unofficial name of the "Three Congruent Glaciers." They originate from very similar firn basins and flow along nearly identical parallel valleys. The one on the left has recently advanced. Its thick, bulging terminus overrides debris in the valley floor, while the thickening edges accumulate a rock mantle from the adjacent valley walls. The center glacier also advanced, but now recedes and has been doing so for a number of years. The rock mantle along each side has been deposited as *lateral moraines*, ridges of unconsolidated material parallel to the glacier flanks. The third "Congruent Glacier" appears to have made a very recent advance from which it has not yet begun to retreat. Faint trimlines considerably higher up the valley walls than the present ice levels indicate that all three glaciers have at one time advanced much farther than their positions today.

38. Different stages of advance and retreat, "Three Congruent Glaciers," St. Elias Mountains

In Figure 39, the north branch of the Tyeen Glacier in the Fairweather Range has advanced rapidly out of its accumulation basin and onto the surface of the main glacier. The exceptionally broken and crevassed surface indicates very rapid ice motion.

39. Rapid glacier advance (surge), Tyeen Glacier, north branch, Fairweather Range

A peculiar type of glacier activity is illustrated in Figure 40a and b. The steep firn field in Sunset Amphitheater, the major accumulation zone of the South Mowich Glacier on Mount Rainier, discharged a vigorous flow of ice in 1960, as shown in Figure 40a. In fact, this steep area of the glacier had turned into an active icefall right up to the bergschrund. Badly broken ice flowed through the narrow breach in a cliff band onto the lower glacier. Two years later, as shown in Figure 40b, the situation had changed drastically: the firn field was now quiescent, and the ice flow through the breach nearly dried up. The South Mowich Glacier exhibits behavior not uncommon to many very steep glaciers. They tend to build up accumulation, reach a critical level, discharge a sudden flow of ice, then return to a quiescent state while accumulation resumes again. The "Spillway Glacier" in the North Cascades (Fig. 41a and b) is another such example.

a

b

40. Build-up and discharge of small, steep glacier, South Mowich Glacier, Mount Rainier, Cascade Range, (a) 1960, (b) 1962

41. Build-up and discharge of small, steep glacier, "Spillway Glacier," North Cascades, (a) 1959, (b) 1960

Changes of a different nature in appearance of a glacier are illustrated in the three photographs of Figure 42. A small glacier on the upper flank of the Aiguille d'Argentière in the French Alps presented a thick layer of ice with a steep frontal cliff in 1899. By 1942 this thick layer and its cliff had entirely disappeared, presumably removed by a change in mass balance which favored ablation more than accumulation. After the next sixteen years not only had this glacier recession ended, but the formation of thick ice cliffs had far exceeded that in 1899. The exact combination of factors affecting these changes is difficult to ascertain, for a glacier like this one hanging on a steep mountainside is strongly influenced by such agents of accumulation and ablation as wind-drifted snow (which depends on storm directions) and both snow and ice avalanching. While the main part of the little glacier almost disappeared by 1942, the rocks below it, prominently visible in the 1899 photograph, are completely blanketed by a smooth layer of snow and ice in the later photograph. Apparently the agents determining the mass balance shifted in a fashion that favored more accumulation on the lower slope and less above. By 1958 a general net accumulation regime had been well established. A prominent bergschrund runs along the foot of the cliffs and steep snow slopes in all three photographs, rising above the active ice of another small glacier mass at the right side of the scene. This bergschrund has remained remarkably stable over a period of fifty-nine years, in marked contrast to the behavior of the glacier.

42. Aiguille d'Argentière, French Alps, (a) 1899, (b) 1942, (c) 1958

43. Advancing glaciers, Roosevelt and Coleman glaciers, Mount Baker, North Cascades

44. Advancing terminus, Coleman Glacier, Mount Baker, North Cascades

Advancing glaciers are in the minority today, although they were far more common in the recent past when a period of cool, wet climate culminated in the "neoglacial" advances of the seventeenth and eighteenth centuries. But even now local climate variations are causing a number of glaciers to advance. Glaciers in the North Cascades of Washington State, after many years of retreat, experienced a period of glacier rejuvenation in the decade beginning in 1945. Glacier advance, shown in Figure 43, was first observed in the Coleman Glacier on the north side of Mount Baker. Later its companion glacier, the Roosevelt, fed from the same accumulation area, also began to advance. Advance of the Coleman Glacier has continued, although the rejuvenation ended for most North Cascades glaciers during the exceptionally long and hot summer of 1958. Since the picture shown in Figure 43 was taken, the Coleman Glacier (in the center) has continued to advance across the stream coming from the Roosevelt Glacier. Figure 44 shows the Coleman Glacier terminus from the ground, the advancing ice a striking contrast to stagnant, retreating termini like the one in Figure 32. Shear planes break the ice wall as the glacier front is thrust forward, while debris picked up by the ice is released by melt at the exposed face and slides down, constantly reforming small moraines of rubble at the ice edge.

45. *Lateral moraines and ground moraine, Easton Glacier, Mount Baker, North Cascades*

Moraines

Moraines have already been mentioned during the discussion of glacier advance and retreat. These debris accumulations are the product of glacier erosion and flow. Their occurrence in a landscape is prominent evidence that glaciers have been present at some time in the past. Moraines associated with present-day glaciers give a glimpse of the way in which similar features were formed on a much greater scale during the Ice Ages.

Lateral moraines are sharp-crested ridges of rubble that mark a valley glacier's flanks in its ablation zone. In this age of generally receding glaciers, they are left standing to mark a glacier's former extent (Fig. 45); between them is found the *ground moraine*, the mantle of debris deposited on the glacier bed. If glacier retreat is interrupted by temporary pauses, or if slight readvances occur, then other sets of lateral moraines, like those displayed in Figure 46, may be built parallel to the first.

46. *Double lateral moraines, Ladd Glacier, Mount Hood, Cascade Range*

47. Surface moraine, Red Glacier, Mount Iliamna, Chigmit Mountains

Ablation moraine sometimes blankets the lower reaches of a glacier. It is produced as melt concentrates at the surface debris that has been gathered and incorporated into the ice during its journey. If this rock blanket becomes thick enough, it insulates the ice and thus severely inhibits further ablation. The Red Glacier in Figure 47 has developed such a thick surface moraine that extensive areas of vegetation have taken root, a development possible only on stagnant glaciers with little movement.

48. Medial moraines, Kaskawulsh Glacier, St. Elias Mountains

49. *Medial moraines, "Speel Glacier," southeast Alaska, Coast Mountains*

One of the earth's most fascinating landscape patterns is the striped appearance which *medial moraines* lend to large glaciers like those shown in Figures 48 and 49. These long, narrow ribbons of rock originate where two ice streams join together, carrying the debris eroded from the glacier margins or deposited there by rockfall into a common line that demarcates the contribution of the two tributaries. If several glaciers join to form a large valley ice stream, each may contribute a medial moraine at the juncture. Volume of ice flow from each branch is shown by spacing of the medial moraines; note the small contributed flow and closely spaced moraines in the center of Figure 49.

Even among the moraine-rich glaciers of Alaska, the Yentna Glacier is distinctive. Its medial moraines originate from different geological structures, each with its own peculiar rock color. The result is a set of parallel medial moraines of different colors side by side, a difference so striking it is obvious even in a black-and-white photograph such as Figure 50.

Figure 51 shows the Malaspina, a huge (2,400-square-kilometer) piedmont glacier built by discharge of the Seward Glacier and various smaller valley glaciers onto the coastal plain of south central Alaska. A complex flow pattern, indicating periodic oscillation of ice discharge from the Seward Glacier, has built an intricate pattern of folded medial moraines as this sheet of ice spreads out across the plain. Medial moraines extend deep below the surface, hence more debris is concentrated at the surface as the ice is removed by melt. This explains why the moraine pattern, faint at the center of the Malaspina Glacier, gets stronger as the ice flows toward the edge. No one aerial photo presently shows the whole extent of this glacier. The sketch map of the moraine pattern shown in Figure 52 was pieced together from several dozen air photos.

50. *Varicolored medial moraines, Yentna Glacier. Alaska Range*

51. *Distorted piedmont moraines, Malaspina Glacier, St. Elias Mountains*

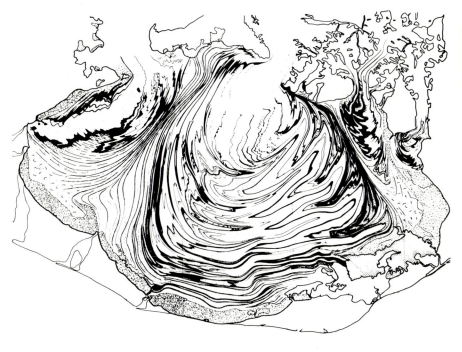

52. *Moraine patterns of the Malaspina Glacier*

53. Looped moraines and advancing glacier, Yanert Glacier, Alaska Range

Glacier Surges

The Yanert Glacier (Figs. 53 and 54) offers a different picture of medial moraines. Twisted and looped into sworls, these moraines are more reminiscent of a chocolate marble cake than of orderly highways (Figs. 48, 49, and 50). Such contorted moraines, peculiar to a few large glaciers, are the product of glacier *surges*. The steep face of the terminus and extensive crevassing show that the Yanert was vigorously advancing in 1942. The advance of a surging glacier occurs on a different scale from that of a glacier which slowly grows in response to a climate change. The speed of terminus advance may reach several kilometers per year for short periods. The unusual features of glacier surges have been identified only recently as a result of extensive aerial reconnaissance. Their exact cause is still a mystery, but the physical characteristics of over two hundred surging glaciers have been observed.

After a long interval (fifteen to one hundred years) when the lower part of a glacier is relatively stagnant, a wave of ice from the upper part abruptly begins to move downvalley. The surface of the glacier is chaotically broken by the progressing wave, and ice may be displaced as much as 10 percent of the glacier length in a single year. The total glacier volume is not augmented; ice for the advancing terminus is furnished by a lowering of surface level farther up the glacier. The stagnant ice in the lower part is activated, thickened, and pushed ahead by the wave. Only in exceptional cases does the glacier advance beyond its former maximum limit.

54. Moraine loops caused by tributary surges, Yanert Glacier, Alaska Range

The active period of a surge generally does not exceed three years regardless of the size of the glacier. Surges occur repeatedly in the same glacier, although they are not always of the same magnitude. There is evidence that the largest displacements of ice occur at long, periodic intervals. Conterminous glaciers may not surge at the same time, and their frequency of surging may differ. There is little relation between glacier size and surge activity; small glaciers only a few kilometers long as well as large ones over one hundred kilometers long may exhibit this phenomenon. On the other hand, surging glaciers are found only in well-defined areas. This limited distribution strongly suggests that some external factor such as unusual bedrock roughness, permeability, or temperature is associated with the ability of a glacier to surge.

The surge of the Muldrow Glacier during the winter of 1956–57 is one of the best documented. Recognizable surface features were displaced downvalley as much as 6 kilometers, while the surface ice level in the upper part of the main valley dropped 60 meters. Figures 55 and 56 show the stranded ice margins (arrows) left by this lowering of the glacier surface. The net ice volume transferred by this surge was about 3.7 cubic kilometers.

55. Stranded marginal ice following glacial surge, Muldrow Glacier, Alaska Range

56. *Close-up of stranded ice following surge, Muldrow Glacier, Alaska Range*

57. Looped medial moraine resulting from periodic surges, Black Rapids Glacier, Alaska Range

The Black Rapids Glacier (Fig. 57) surged ahead 5 kilometers in three months during the winter of 1936–37. The looped medial moraines, similar to those of the Yanert Glacier (Fig. 53), result from a steady discharge from a tributary into a main ice stream which alternates between surge and stagnation.

The Walsh Glacier is another large valley glacier in the course of a surge at the time Figure 58 was taken in 1963. Note the looped moraines and extremely broken surface, both characteristic of surging glaciers.

58. Large surging glacier, Walsh Glacier, St. Elias Mountains

59. *The generation of looped and folded moraine patterns by periodic surges of a valley glacier with steady state tributary, Susitna Glacier, Alaska Range, (a) 1941, (b) 1966*

The most famous set of contorted medial moraines in Alaska is that of the Susitna Glacier (Fig. 59). The 1941 photograph by Bradford Washburn (Fig. 59a) has been widely published. Figure 59b, taken twenty-five years later, shows the moraine pattern completely changed. In 1941 the left-hand tributary was providing the major part of the ice flow, almost pinching off the stagnant main glacier. In 1952 a surge of the latter displaced the ice 4 kilometers downvalley. Since the surge, a new loop in the moraine has started to form as the tributary pushes into the again stagnant main stream. These changes can be more clearly seen in the sketch maps of Figure 60.

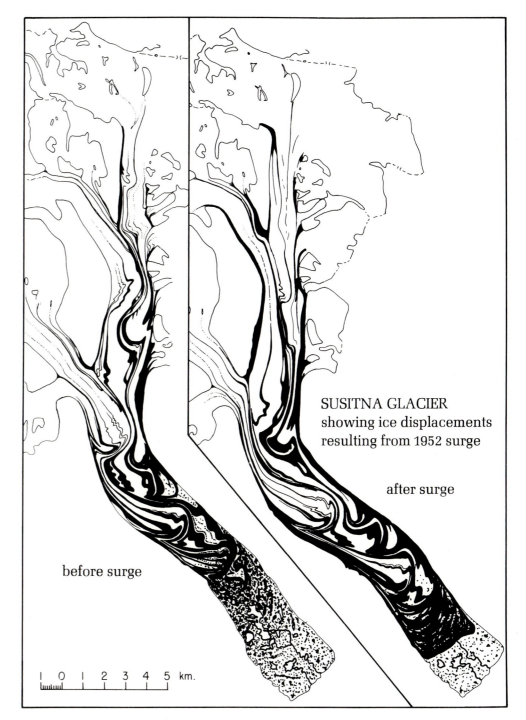

SUSITNA GLACIER
showing ice displacements
resulting from 1952 surge

after surge

before surge

1 0 1 2 3 4 5 km.

60. Sketch maps of Susitna Glacier surge and moraine patterns

61. *Wave ogives, Yentna Glacier, Alaska Range*

Ogives

Some of the most striking patterns in nature are presented by the regularly spaced arcs that appear on certain glaciers. These *ogives* occur as two types: wave ogives, or undulations in the glacier surface; and band ogives, or alternate dark and light bands on a smooth ice surface. Figures 61 and 62 illustrate these two basic types, respectively. Figures 63, 64, and 65 are other well-known examples of band ogives.

62. Band ogives, Mer de Glace, French Alps

63. *Close-spaced band ogives, Battle Glacier, southeast Alaska, Coast Mountains*

64. Band ogives, East Twin Glacier, southeast Alaska, Coast Mountains

65. Structured band ogives, Gilkey Glacier, southeast Alaska, Coast Mountains

66. *Generating icefall for Gilkey Glacier ogives, Vaughn Lewis Glacier, southeast Alaska, Coast Mountains*

67. *Persistent wave ogives, Trimble Glacier North Branch, Alaska Range*

Ogives originate when a glacier flows down a steep, narrow icefall. Not all icefalls produce ogives, and very few produce the very pronounced ogives illustrated here. Several theories have been advanced to explain how these features form, but no one theory is yet general enough to account for all the observed varieties. It is known that glacier flow velocity is much higher in an icefall than in the rest of the glacier. One plausible theory suggests that this local acceleration stretches out the ice and exposes more area per unit volume as it passes through the icefall. The exposed area is further increased by crevassing and fracturing. The ice in transit during the summer melt season thus suffers more than its normal share of ablation, while that passing over the icefall in winter is protected by a mantle of winter snow. The net result is annual fluctuation in the thickness of ice emerging at the bottom of the icefall even when flow velocity is constant with time. Waves develop further as these variations in thickness are compressed in the slow-moving ice below an icefall, as shown in Figure 66. In some cases an annual variation in flow velocity may add to the process as well. The alternate dark and light ogive bands are thought to originate in a similar fashion, the dark bands representing ice that has passed through the icefall in summer and accumulated extra surface dirt which is later concentrated by ablation. The Gilkey Glacier (Fig. 65) has an especially clear set of band ogives with fine structure visible in the bands. As in the case of many band ogives, these result from wave ogives formed at the foot of the generating icefall, in this case that of the tributary Vaughn Lewis Glacier (Fig. 66).

It is far from clear why some ogive systems such as those in Figures 67 and 68 persist as waves a long distance downglacier, while others (Figs. 63 and 64) quickly degenerate from waves into bands. The whole problem of ogive formation is far from solved, and future investigations may well lead to views different from those outlined above. It should be noted, for instance, that wave ogives are actually a rather superficial feature of the glacier ice; their amplitude is only a very small percentage of the total ice depth at the foot of an active icefall. Why do some icefalls generate ogives while others in similar circumstances do not? The answer to this question is no doubt closely connected with some yet-unresolved character of glacier flow.

68. *Wave ogives buried under firn, Grand Pacific Glacier, Fairweather Range*

69. *Wave ogives and generating icefall, Gerstle Glacier, Alaska Range*

70. Close-up of Gerstle Glacier ogives, (a) 1960, (b) 1961

The Gerstle Glacier ogive system is shown in Figure 69. That ogives have annual periodicity can be clearly demonstrated from Figure 70, with two photographs of the Gerstle Glacier icefall taken almost exactly a year apart. Debris features on the ice permit the same ogive waves to be identified in both pictures, and these have been numbered. In 1960 the crest of wave No. 6 was just forming at the icefall foot. A year later the ogive system had moved downglacier approximately one wave length in respect to adjacent landmarks; wave No. 6 had formed completely, followed by a trough; while the crest of No. 7 was just starting to form.

71. *Band ogives plus accumulation layering, Sherman Glacier, Chugach Mountains*

Ogives are a product of glacier flow dynamics and should be distinguished from the banding or layering that results from stratification of annual accumulation such as displayed in Figure 9. There was considerable confusion on this point among early investigators in the Alps, but the Scottish geologist, J. D. Forbes, correctly pointed out the difference as early as 1843, and band ogives are sometimes called "Forbes Bands" to this day. These two different patterns occur side by side on a single glacier in Figure 71. The faint ogives have a characteristic arcuate pattern, while the irregular bands delineate the uneven intersection of an ablation surface with the accumulation strata.

In contrast to uniform, closely spaced normal ogives are the remarkable "false ogives" on the Margerie Glacier in the Fairweather Range, southeastern Alaska (Fig. 72). These do not originate from an icefall, and their wide spacing makes an annual periodicity implausible. In fact, faint annual ogives can be detected between the second and third heavy bands from the bottom of the figure. The Fairweathers present some of the steepest and most inhospitable terrain in North America, and large rock avalanches are frequent. Careful examination of photographs taken in different years shows that the dark bands on the Margerie Glacier represent debris from periodic rock avalanches which descended onto the glacier every ten years or so from the peak inside the curve where the glacier disappears from view in Figure 72. It is remarkable that such similar rock slides have occurred at such uniform intervals.

72. False ogives (debris from rock avalanches), Margerie Glacier, Fairweather Range

73. Meltwater stream of glacier surface, Blue Glacier, Olympic Mountains

Meltwater on Glaciers

Glaciers everywhere except in the high polar regions experience a substantial amount of melt on at least part of their surface each summer. Meltwater like that shown in Figure 73 makes its way over, through, and beneath the glacier, eventually appearing at the terminus as a stream. The discharge of this stream fluctuates widely during the year. It also fluctuates widely during each day as the amount of meltwater released varies with daytime melt and nighttime freezing. The pattern of discharge from a glacier-fed stream can be just the opposite of that from a river draining a normal, ice-free valley. The latter has its maximum flow during winter months when rainfall is high, or during the spring snow-melt period. It is lowest late in the summer in regions where summer rainfall is slight. A glacier stream, on the other hand, discharges little water in winter because its source, the melt of ice, is cut off by freezing temperatures and by an insulating blanket of winter snow. Such a stream flows highest in the summer when glacier melt is at its peak. The presence of even small glaciers in a river basin thus tends to stabilize the total river flow, furnishing a maximum discharge of water just when other sources are at their lowest.

Streams often run for long distances over the surface of a glacier before finding their way down to its bed or off its margins. These rivers may develop the same dendritic patterns of tributaries (Fig. 74) as normal rivers running on land. On large glaciers these surface streams sometimes reach a very large size in summer, cutting deep channels with steep and dangerously slippery walls of ice. Few such streams follow the surface of the ice all the way to the terminus. Most eventually flow into a subglacial stream channel which leads to the glacier bed by a sometimes intricate passageway. The hole where the meltwater stream plunges into the glacier depths is called a *moulin*, or *glacier mill*, a name stemming from the rumbling sound of the water, like that of an old-fashioned water-powered mill. Figure 75 shows a moulin with a vigorous stream plunging in at the bottom of the picture.

74. Dendrite stream pattern on glacier surface, Black Rapids Glacier, Alaska Range

75. *Moulin, Glacier du Miage, Mont Blanc, French Alps*

76. *Snow swamp, Greenland Ice Sheet*

On the exposed blue ice of glacier ablation zones, meltwater flows on the surface because of the impervious character of the ice. In the accumulation zone, where the glacier surface layers consist of porous firn, the meltwater does not stand in open pools or streams at the surface except where it is prevented from draining by an ice layer. Sometimes a water table exists only a short distance below the surface, but is not visible. The apparently smooth snow surface conceals slush below. This condition can generate unpleasant surprises for the unwary glacier traveler. In Figure 76, a tracked vehicle has broken into one of these *snow swamps*. If the snow becomes sufficiently saturated on a gentle but perceptible slope, it can break loose in a "slush avalanche" which sweeps down the glacier at high speed.

77. Potholes, Black Rapids Glacier, Alaska Range

78. Potholes, Black Rapids Glacier, Alaska Range

The peculiar potholes on the surface of the Black Rapids Glacier and its tributary, shown in Figures 77 and 78, probably owe their origin to perennial meltwater ponds on the surface of nearly stagnant, crevasse-free ice. Somewhat similar features on a much larger scale, called *ice dolines*, have been observed in Antarctica. Their formation has been postulated to be caused by depressions in the ice which fill with summer meltwater, then freeze over. Later glacier motion opens a subglacial passage which drains away the water, and the ice "lid" then collapses.

79. Accumulation zone crevasses, Blue Glacier, Olympic Mountains

Surface Details

Most of the glacier features illustrated up to this point have been photographed from the air. This viewpoint is often required to examine the larger aspects of glacier ice, but there are many details of glacier ice that are better seen from the ground.

The most frequent glacier traveler is the mountain climber. Normally he will avoid shattered glacier surfaces (Fig. 23), for ice broken to this extent is virtually impassable. An orderly arrangement of crevasses like that in Figure 79 is the sort of glacier terrain a mountaineer can traverse without difficulty. This picture shows an accumulation zone in a maritime climate of heavy snowfall. The crevasses are large and well developed, but spanned by numerous firn bridges and far enough apart to permit some freedom of movement. The accumulation zone of a glacier is a region of extending flow where the ice velocity increases downstream. This means that the surface is being stretched. Most of this stretch takes place by the widening of crevasses, to which is added the effects of melting. By late summer even a simple crevassed area like this may become nearly impassable through the collapse of snow bridges and extension of open crevasses clear across the glacier. The mountaineer then must thread his way along precarious routes like the one in Figure 80 (see also Fig. 19).

80. Ridge between firn crevasses, Blue Glacier, Olympic Mountains

81. *Ablation zone crevasse, Chogo-Lungma Glacier, Karakoram Himalayas*

82. *Inside of a crevasse, Blue Glacier, Olympic Mountains*

By all odds the most popular view of a crevasse is from above, looking in. The aerial photographer can view these pitfalls with impunity, and the wise mountaineer takes great care to avoid falling into glacial chasms like the one in Figure 81, a good example of a large ablation-zone crevasse that has been extensively eroded by melt. Those who do happen to find themselves plunged into such a crevasse (hopefully attached to a resilient climbing rope) rarely give serious thought to photography in their anxiety to regain the surface. Figure 82 is the view from the bottom of a crevasse, looking up. It was obtained under unusual circumstances. A recent scientific project on the Blue Glacier, located in Olympic National Park, involved digging a tunnel into an icefall to study sliding of the glacier over its bed. This tunnel intersected the bottom of a crevasse, permitting easy access (and egress) by which the photographer could obtain the picture at leisure. The intricate passages and rough surface are typical of ice that has been sheared and then exposed to melt, probably in this case by circulation of warm air from the surface.

The photograph in Figure 83, taken on the ablation zone of the Great Aletsch Glacier in Switzerland, illustrates a number of different glacier features. The rock debris mantling part of the ice is a medial moraine. Such rock material is distributed vertically through the glacier ice and becomes concentrated at the surface as ablation lowers the ice surface exposing the rocks. The dark area on the wall of the crevasse that traverses the foreground shows where the ice is filled with rock debris. The thin layer of dirt and rocks concentrated on the surface is darker than the surrounding ice, absorbs more heat from the sun, and contributes to faster ice melt. For this reason the debris-mantled surface is lower than that of clean ice. If the debris layer is thick enough, however, it insulates the ice from the extra heat and inhibits ice ablation. Ice so shielded stands above the level of the normally melting ice. If a rock is large enough to inhibit ice melt and shaped so that it does not readily tumble from the resulting pedestal, it becomes elevated above the surface as a *glacier table*.

83. Glacier table and foliated ice, Aletsch Glacier, Switzerland

On the left side of Figure 83, several vertical stripes on the wall of the crevasse are obvious extensions of similar stripes on the glacier surface. These stripes define the edges of planes of *foliation*. They are visible at the surface because these planes contain either slight concentrations of dirt, or ice whose capacity to absorb solar radiation is different from that of the surrounding ice. Either case leads to local alteration of melt and consequent irregularities in the glacier surface. Foliation is the layering in glacier ice produced by rapid deformation which causes the ice to recrystallize along certain planes. Hence it is a product of glacier flow that is completely unrelated to accumulation layering like that in Figures 10 and 11.

Not all the surface features on a glacier derive from water, rock, and ice. Surprisingly, living creatures make their permanent home on the glacier surface. The *ice worm* has been so widely celebrated in songs and stories of the northland that many regard him a myth, to be classed with the griffin and the unicorn. Contrary to popular supposition, he is not the product of imagination, he does not bear suspicious resemblance to a piece of cooked spaghetti with ink spots for eyes, nor is his normal habitat the bottom of a cocktail glass. The ice worm is a common inhabitant of the glaciers and permanent snowfields found in the temperate, maritime climate of the North Pacific Coast. He is, in fact, an annelid worm belonging to the class Oligochaeta, and thus a relative of the common earthworm.

This peculiar worm thickly populates the glacier snow surfaces of early summer, and on a clear summer evening many hundreds to the square meter can be counted over large areas of a glacier. The worms in Figure 84 were not gathered together for a picture; they were photographed in their natural abundance on the summer snows of the Blue Glacier.

85. Sun cups, Taku Glacier, southeast Alaska, Coast Mountains

84. Ice worms, Blue Glacier, Olympic Mountains

The ice worm probably has one of the most remarkable temperature environments of any creature on earth. His entire life cycle is spent very close to the freezing point of water, subsisting on the pollen or algae incorporated in the snow by wind and concentrated at the surface by ablation. During the day the ice worm retreats 30 centimeters or more below the surface to avoid heat from the sun, but he comes out in the evening or during cloudy weather, presumably to search for food. As soon as radiation cooling on a clear night starts to freeze liquid water at the snow surface, the ice worm retreats once more, staying beneath the chilled nocturnal crust of frozen snow. A similar cycle happens on an annual basis. Each autumn as the chill (subfreezing temperature) penetrates deeper and deeper into the firn, the ice worm retreats farther into the glacier. When spring melt dissipates the winter chill, it begins working up to the surface through all the accumulated snow, to appear in May or June. Such a life cycle can occur only on permanent snow and ice, and thus the ice worm is never found on transient winter snowfields. In fact, its presence or absence on a particular part of the glacier surface can clearly delineate the margin separating glacier ice from annual snowfields.

When snow or firn melts, the process of ablation never proceeds with complete uniformity. Slight irregularities in the snow structure, or local variations in melt, or both, soon produce a rough surface. There are several processes that can cause the roughness to increase as melt continues. Whenever this happens, the snow or firn surface soon develops a pattern of *sun cups* like those in Figure 85. These hollows in the snow surface can range from shallow ripples only a centimeter or so deep to patterns with a relief of half a meter or more (Fig. 86).

87. *Nieve penitente, north slope of Cerro Marmolejo Norte, Chile*

88. *Highly developed nieve penitente, Cerro Negro, Chile*

Sun cups begin to form whenever weather conditions favor evaporation at the high points of the snow surface and melt in the hollows. There are definite conditions of temperature and humidity under which this is possible, with the additional requirement of an extra source of heat other than that from warm air. Normally this extra heat can come only from sunshine; hence the name "sun cups." About $7\frac{1}{2}$ times as much heat is required to evaporate a given mass of ice (or snow) as to melt it. Even if more heat reaches the high points of a rough snow surface than reaches the hollows, this may actually cause reduced ablation at the high points when enough evaporation takes place there. Thus the hollows ablate more rapidly, and the sun cups grow deeper.

Once the snow surface has been transformed from rough irregularities to regular sun cups, another mechanism begins which can continue even when favorable atmospheric conditions are not present. Percolating meltwater flows from the peaks down the sides of the cups to the lowest point in the center, then drains away through the snow. The finest particles of dirt concentrated at the surface by ablation are carried by this meltwater and congregate at the bottoms of the cups. The bottoms of these hollows grow darker than the crests, absorb more heat from the sun, and experience increased melt which further deepens the cups.

89. *Ice pillars, Khumbu Glacier, Mount Everest*

In higher latitudes where most glaciers are found, the sun seldom or never stands directly overhead. When not at the zenith, the sun shines more directly on one side of the sun cup than on the other, causing more melt on one side of the cup. This causes a field of sun cups to migrate as the cups deepen and form. In the Northern Hemisphere this migration is toward the north. Under favorable circumstances it can amount to several centimeters a day.

The most extreme weather conditions favoring sun cup development are found at very high altitudes in the tropics. Here the air is cold and usually very dry, the reduced atmospheric density augments evaporation, and intense heat is available from the sun. In such circumstances, sun cups keep on growing for weeks or months. The cups soon intersect, leaving parts of the ridges between standing as pinnacles. The greater the relief, the greater is the difference in ablation between pinnacle and hollow. The ultimate result is a field of *nieve penitente*. This name, Spanish for "penitent snow," originated in South America where the phenomenon is frequently seen in the high Andes. With the advent of warm weather, the tips of the snow pinnacles often droop or bend over, giving the appearance of ranks of hooded monks with their head bowed in prayer— hence the term "penitent." In Figure 87, the formation of hollows at the bases of the pinnacles has proceeded right through the winter snow cover to the ground, leaving in some places arrays of isolated pinnacles. The area in the foreground shows strong formation of nieve penitente in snow; the background exhibits a similar formation in glacier ice. Figure 88 is an extreme example of nieve penitente. Layering of the snow is plainly visible in the sides of the pinnacles.

High-altitude, low-latitude glaciers mantled with rock debris (ablation moraine) sometimes develop surface structures analogous to nieve penitente. These ice pinnacles are built in part or sustained by the evaporation-melt processes described above, but they also experience a greatly enhanced effect from the difference in solar heat absorbed by the dark debris mantle and the white ice surfaces. Many of these pinnacles probably begin as the supports for glacier tables (Fig. 83). The cap rock subsequently topples off, leaving pillars like those in Figure 89. Subsequent ablation then alters their shape to the triangular pinnacles of Figure 90 (see also foreground in Fig. 12). Under favorable circumstances these pinnacles may continue to grow to a very large size. They do not, of course, grow in the sense of getting higher at the top. The debris-covered ice melts downward much faster than clean, white pinnacles, so they are left behind as the remnants of a rapidly eroded glacier surface. The giant pinnacles like those in Figure 91 were named *ice ships* when they were first discovered on the Baltoro Glacier in the Karakoram Himalayas.

91. *Large ice ships, Khumbu Glacier, Mount Everest*

90. *Small ice ships, Khumbu Glacier, Mount Everest*

92. *Tidewater glacier, Johns Hopkins Glacier, Fairweather Range*

Tidewater Glaciers

In polar regions, and in coastal ranges with heavy winter precipitation and sufficiently low mean annual temperature, glaciers reach low elevations before summer melt balances the flow of ice. Many glaciers flow from accumulation areas so large that even at sea level the volume of ice melt each year is much less than the volume of ice carried down from higher elevations. Such glaciers which reach the sea are called *tidewater glaciers*, like the Johns Hopkins Glacier shown in Figure 92. Here a different process of ablation is added to the normal one of melt from sunlight and warm air. This is the calving of *icebergs*, which are carried away by ocean currents eventually to be melted at sea. Calving is a continuous process in an active glacier as the ice is shoved forward by glacier flow and constantly eroded by the water. Figure 93 shows a serac that has just toppled over from the glacier front to form an iceberg, in this case in a freshwater lake. These falling bodies of ice, and the large waves they generally produce, can make close approach to a calving glacier extremely dangerous.

93. Calving iceberg, Miles Glacier, Copper River Valley, Alaska

94. *Retreating tidewater glacier, Guyot Glacier, south central Alaska, (a) 1938, (b) 1963*

a   b

Tidewater glaciers sometimes retreat with spectacular rapidity. In shallow channels this process is restricted, but the breakup is rapid when the channel is deep and broad. Figure 94 illustrates tidewater recession of the Guyot Glacier, amounting to about 165 square kilometers of ice in twenty-five years. Most of the loss occurred in only six years when the retreating glacier opened a large, very deep tidal basin.

The Guyot Glacier exhibited rapid iceberg discharge and retreat when the channel was not deep enough to float the thick glacier ice that existed at that time. Another example of this situation is found in Figure 95. Only a short time before the picture was taken, the Dawes Glacier had extended two kilometers farther down channel to the rocky promontory at lower left. This long stretch of the glacier broke up into innumerable icebergs in a few months. This was first noticed when large numbers of icebergs suddenly appeared in the shipping lanes of the Inside Passage between Juneau and Ketchikan. As in the case of the Guyot Glacier, the height of the Dawes ice cliff above the water was too great for the ice to have been afloat in the depth of water present; hence some other mechanism besides simple flotation must have caused the breakup. Where the valley bottom is far below sea level, the rapid retreat of ice from one promontory or constriction in the channel to the next has been repeatedly noted. Headlands and channel constrictions provide anchor points where the glacier terminus often maintains a tenuous stability, sometimes for several years. Once the ice retreats from these points, breakup is catastrophic, and the glacier shrinks rapidly until the next constriction in the channel is reached. When tidewater glaciers retreat in this fashion, a coincident acceleration of flow often takes place, judged to be due to the rapid steepening in longitudinal gradient. Under these conditions, ice flow averaging nearly 20 meters per day in Guyot Glacier was recorded over a three-month period in 1948.

95. Tidewater glacier and ice-choked fjord, Dawes Glacier, southeast Alaska, Coast Mountains

The deep, steep-walled valleys produced by glacial erosion are called *fjords* when they are inundated by the sea. This most frequently results from a glacier eroding its bed far below sea level in these narrow, mountainous canyons. Fjords are typically much deeper than the coastal waters with which they merge. Many previously carved fjords are occupied in part by present-day tidewater glaciers such as Meares Glacier, shown in Figure 96. When the ice retreats, it leaves behind a water-filled fjord like Tracy Arm, as in Figure 97, which presents a remarkably accurate picture of what the Meares Glacier fjord would look like if the glacier were not present.

96. Glacier-filled fjord, Meares Glacier, Chugach Mountains

97. *Ice-free fjord, Tracy Arm, southeast Alaska, Coast Mountains*

98. *Ice-blanketed volcanic cone, Mount Rainier, Cascade Range*

Glaciers and Volcanoes

The close association of glaciers and volcanoes produces a number of unusual glacier phenomena. Such association is known for the prominent volcanoes of the Pacific Northwest (Figs. 98 and 99), but probably is best documented for the glaciers and volcanoes of Iceland. Glaciers that blanket the upper flanks of steep volcanic cones produce an especially rapid erosion of the soft and friable layers of lava, pumice, and volcanic ash. Wind transport of dust derived from these materials quickly causes the accumulation of a dark surface layer which is concentrated by summer melt and enhances absorption of solar heat. Local acceleration of snow and ice melt takes place wherever fumaroles (steam vents) discharge steam and hot volcanic gases. Such a vent discharges a prominent plume of vapor among the glaciers of Mount Chiginagak in Figure 100.

If a summit crater, or caldera, exists, it is an excellent catchment for snow. This may provide the zone of principal nourishment for several glaciers like those on Mount Rainier (Fig. 98). The summit crater of Mount St. Helens generates a peculiar glacier which flows through a breach in its eastern wall. The extreme narrowness of this ice stream, shown in Figure 101, has earned it the descriptive name of Shoestring Glacier.

99. *Volcanic cone with several small glaciers, Mount St. Helens, Cascade Range*

100. Glacierized volcano with active fumarole, Mount Chiginagak, Aleutian Range

The hydrologic cycle of snow accumulation, ice flow, melt, and runoff is a regular one for most glaciers. The meltwater discharge follows a pattern of minimum flow in winter and a maximum flow in midsummer during the height of the melt season. In certain glaciers this pattern may be interrupted by the sudden, brief discharge of very large quantities of water. Such an abnormally large discharge may be irregular and take place at widely scattered intervals, or the occurrence may be annual. Enormous quantities of water released in a very short time by glaciers often cause catastrophic floods in the valley below. Close association of glaciers and volcanoes favors this type of flood in Iceland, where it is sufficiently common to have been given a name, *jökulhlaup*. This term is now used elsewhere to describe catastrophic glacial floods.

The jökulhlaup originates from large bodies of meltwater impounded by the glacier along its margin, beneath, or even within, the ice. The outlet for these lakes or caverns can be closed by ice flow when the reservoir is depleted. The water level, rising with advent of the ablation season, heavy rainstorms, or through melt caused by volcanic heat, fills the reservoir. The initial breakthrough starts with a trickle of water through, over, or under the ice. If the impounded water is sufficiently deep, the hydrostatic pressure of the water can also exert sufficient force on ice at the bottom to deform it by plastic flow. The "leak" may also start through natural cracks, or simply by the water overtopping the ice. Once the discharge starts, flow water even slightly warmer than 0°C rapidly melts and enlarges the channel. The larger the channel becomes, the more rapidly the water escapes, providing additional heat to melt more ice. Such a sequence of events leads to an exponential rise in water discharge; the result is a sudden "dumping" of the impounded water in a matter of hours.

101. Shoestring Glacier originating in volcanic summit crater, Mount St. Helens, Cascade Range

The Tulsequah Glacier, one of the outlet glaciers of the Juneau Ice Field, is famous for its annual jökulhlaup, which originates from glacier-dammed Tulsequah Lake, located along its west margin. Figure 102 shows the lower reaches of this glacier, with the site of the lake indicated by the arrow. This lake occupies a deep valley into which flows a distributary arm of the glacier shown in Figure 103. Each winter and spring this valley fills with water to a depth of about a hundred meters, indicated by a clearly visible strand line. The breakout usually comes in July, when a subglacial passage is opened beneath the glacier. This photograph of the lake was taken in 1961 about a month after the annual breakout; already the runoff from surrounding glaciers has begun to refill the basin.

Less clear is the origin of infrequent, erratic floods from glaciers mantling the steep flanks of volcanic cones. Here there is no obvious reservoir in which large quantities of water can be stored for sudden release, such as the glacier-dammed valley of Tulsequah Lake. Yet at infrequent intervals a destructive flood may emerge without warning from a glacier and sweep down the valley below. The glaciers of Mount Rainier have yielded a number of such floods in historical times, and geological evidence indicates that others have occurred in the past. The Kautz Creek Flood of 1947 was unusually destructive, for it destroyed over 1 kilometer of the glacier, cut a deep new stream channel, deposited mud and rocks 10 meters deep over a highway, and inundated a mature forest. Stratigraphy of the new stream banks disclosed that the Kautz Glacier had previously discharged similar floods. Though the 1947 flood followed an exceptionally heavy rainfall, a sudden release of water by the glacier was obviously involved. Other floods in this area (for example, the Kautz Glacier Flood of 1960) were not associated with rainfall. A more recent example is seen in Figure 104 on the South Tahoma Glacier. A hidden reservoir of water suddenly broke out in midglacier and flowed over the ice, eroding a deep channel (arrow) before the cavern drained and the flood subsided. These floods clearly resulted from the sudden discharge of water stored within or under the glaciers, and it is difficult to conceive how such large englacial or subglacial cavities can exist in an actively flowing ice. It has been suggested that heat suddenly released from within the dormant volcano may produce the water by melting ice at the bed of the glacier.

102. *Tulsequah Glacier, producer of annual jökulhlaups, British Columbia, Coast Mountains*

103. *Glacier-dammed lake, Tulsequah Lake and Glacier, British Columbia, Coast Mountains*

104. *Flood channel on South Tahoma Glacier, Mount Rainier, Cascade Range*

105. *Glacier striations in bedrock, Blue Glacier, Olympic Mountains*

Effects of Glaciers on the Landscape

Glaciers are very active agents of geological erosion; much of the spectacular mountain scenery of the earth has been sculptured by ice. Although ice itself is a very soft material with little abrading power on rock, glaciers carry with them much rock debris embedded in the ice, especially in the sole (see, for instance, Figs. 30 and 31). This rock debris serves as an effective tool with which the ice gouges and grinds the underlying bedrock during its journey down the mountainside. Bedrock exposed by a retreating glacier has imprinted on it the evidence of this erosion in the form of grooves and striations generated by rock particles of varying sizes in the glacier sole. Figure 105 shows such striations in freshly exposed bedrock along the margin of the Blue Glacier. Direction of ice motion was from upper right to lower left. The larger pattern of grooves or undulations is also visible, as well as an area just above the ice axe where a large piece of rock has been detached by the plucking action of the ice. Such plucking is one way the glacier acquires the rock tools for erosion.

One product of rock grinding against rock is *glacier flour,* the extremely fine dust that remains suspended in glacier-fed streams, giving them a milky appearance, and which even remains suspended in lakes fed by such streams. The content of rock powder gives these lakes unusual colors, ranging from muddy chocolate to brown, tan, cream, and even a brilliant emerald green.

106. *Massive rock avalanche resulting from earthquake shaking of cliffs steepened by glacial erosion, Sherman Glacier, Chugach Mountains*

The rapid erosion by glaciers deepens and broadens stream valleys into U-shaped canyons. These canyons and the mountain peaks resulting from ice erosion and severe frost action are steep-walled landforms very different from the smooth-crested, often intricately gullied slopes produced by other weathering agents. If composed of poorly consolidated rock, glacially steepened slopes are unstable and subjected to large-scale catastrophic collapse. Figure 106 shows an example of such a catastrophe. The Alaska Good Friday earthquake of 1964 dislodged many landslides among the glacier-carved peaks of the Chugach Mountains. One of the most spectacular was this one, which blanketed the surface of the Sherman Glacier with a hugh mantle of rock debris. This slide originated from the face of Shattered Peak, just below the towering cumulo-nimbus clouds in the center of the photograph. It flowed at extremely high velocity 6 kilometers over an intervening ridge and then at a very low gradient across the Sherman Glacier. It has been postulated that this remarkable flow was due to the formation of a lubricating air cushion under the rapidly moving debris.

The erosive action of moving ice causes a glacier to cut its bed deeper and deeper, broadening the valley floor and cutting back into the headwall of its accumulation basin to form a *cirque*. The character of the mountain terrain thus developed depends on the structure and quality of the bedrock as well as on the nature of ice flow. Massive, granitic bedrock especially tends to form spectacular mountain scenery under the sculpturing of glacier ice. An early stage of glacial erosion in granite is seen in Figure 107, where scattered cirques have been cut into an elevated tableland, forming a *biscuit-board topography.* When glaciation is prolonged beyond the biscuit-board stage, the cirques and glacier-carved valleys eventually intersect. The jagged ridges that mark these lines of intersection are called *arêtes,* illustrated in Figure 108. On a larger scale, deep, glacier-carved gorges like the gigantic cirque in Figure 109 are also ringed with sharp aretes. When ice has intensively eroded massive granite, the arêtes stand to the sky in dizzying pinnacles and ridges that attract the more daring mountaineers. The Kichatna Mountains in Figure 110 are a good example of this stage of glacial erosion. In regions of heavy snowfall and prolonged glaciation, ice has worn away all except the most durable and resistant rock outcrops, and *horns* are left as the only remnant of a long-vanished upland (Fig. 111).

107. Biscuit-board topography, Wind River Mountains, Wyoming

108. *Serrated arêtes, Eldorado Peak area, North Cascades*

109. *Glacier gorge and arêtes, tributary to Johns Hopkins Glacier, Fairweather Range*

111. *Horn-type mountain, Devil's Thumb, British Columbia, Coast Mountains*

112. *Juneau Ice Field, southeast Alaska, Coast Mountains*

113. *Ice field and nunataks, Bagley Ice Field, south central Alaska*

The effects of glacial erosion (Figs. 107–111) are primarily those of scattered alpine glaciers, although in some cases, as seen in Figure 111, there is definite evidence that the terrain was once inundated by a more extensive ice cover. *Ice sheets* presently cover practically all of Greenland and Antarctica. Comparable ice sheets covered areas in Asia, Europe, and North America during the Pleistocene. Several much smaller *ice fields* exist today. The Juneau Ice Field in southeast Alaska shown in Figure 112 is one of the best-known examples. Here a sea of ice fills the valleys, and only the highest ridges and peaks remain uncovered. These areas of bedrock completely surrounded by ice are called *nunataks,* a word of Greenlandic Eskimo origin. The Bagley Ice Field (Fig. 113) contains row after row of nunataks, each row consisting of the highest peaks of an ice-flooded mountain range.

When a landscape becomes almost completely buried under an ice field, the rough, irregular peaks and ridges are worn away by the ice. Steep-walled valleys occupied by the principal glaciers draining the ice field are smoothed and rounded. When the ice vanishes, a more subdued landscape remains, broken by deep valleys and an occasional high, rocky promontory which as a nunatak escaped inundation. Figure 114 shows such a landscape, suggesting what the terrain of the Juneau Ice Field would look like if the ice melted away.

114. *Formerly ice-covered terrain, British Columbia, Coast Mountains*

115. *Freshly deglacierized terrain, Muir Inlet, Glacier Bay, southeast Alaska*

116. *Glacial till, Jmjia Glacier, Garwhal Himalayas*

117. *Esker, Malaspina Glacier, St. Elias Mountains, Alaska*

Icecaps, generally defined as ice domes from which no mountain peaks emerge, are most frequently found on plateaus or uplands in polar regions.

When glaciers melt, they uncover distinctive glacial deposits from which many features of ice erosion can be deduced. For instance, melting of a remnant of the Muir Glacier, on the east side of Glacier Bay, has exposed an unusual display of glacial geology (Fig. 115). The glacier had recently overridden this area of poorly consolidated sediments, and the flowing ice produced the series of gently rounded, parallel ridges aligned with the direction of flow. These are elongated forms of *drumlins,* or mounds of glacial-sculptured material left behind in an orderly array by a retreating glacier. Running across at right angles to these drumlin ridges are many narrow streaks of debris which stand out sharply, owing to their fresh and uneroded appearance. This debris, which had collected in the bottoms of crevasses, was lowered undisturbed onto the ground as the ice melted. Rapid incision of stream channels into the freshly exposed ground has exposed various former glacier outwash deposits, some containing forest remains, and layers of glacial *till* deposited by previous glacier advances. Glacial till, illustrated in Figure 116, consists of unstratified and largely unsorted rocks, gravel, and finer debris all jumbled together.

An *esker* is a ridge of sediment deposited by a subglacial meltwater stream. It follows the twists and convolutions of the generating stream. Eskers can form only in stagnant ice where the sediment can build up to appreciable depths and be left in place undisturbed when the ice melts. The twisting channel of an esker-building stream is especially well delineated in Figure 117. This esker, near the edge of the Malaspina Glacier (Fig. 51), has the additional peculiarity of being deposited on ice rather than on bedrock. Evidently it was formed by a meltwater stream whose channel did not extend to the glacier bottom. Two rough irregular eskers are also visible in the upper part of Figure 115.

Figure 118 shows an area of terminal moraines from which the generating glacier retreated a century or more ago. Many of the fine details of glacial deposits visible in Figure 115 were no doubt also present here, but have long since been obscured by erosion and concealed by thick vegetation. The outlines of some of the old terminal moraines can still be seen as patterns in the timber growth. Patterns like this are easy to recognize from the air but often difficult to detect on the ground.

118. *Forested moraines, Herbert Glacier, southeast Alaska, Coast Mountains*

119. *Small polar ice field, Axel Heiberg Island, Northwest Territories*

Temperate, Subpolar, and Polar Glaciers

There has been a common feature in the illustrations presented up to this point. All of the glaciers have been *temperate glaciers,* whose ice is very near the pressure melting point throughout except for a surface layer that is chilled below freezing each winter. The mode of flow of temperate glaciers is profoundly affected by the presence of liquid water throughout most of their mass. The ice deforms, fractures, and slides readily, and the regelation process is able to assist sliding of the ice over bedrock. The overall rate of glacier flow depends in part on the amount of liquid water produced by rainfall and surface melting.

Subpolar glaciers, on the other hand, exist perennially at subfreezing temperatures throughout most of their mass. Some liquid water may be present at the surface and in limited channels during the melt season. Glaciers in very cold climates may also be frozen in part to their bed, with the regelation process eliminated as a mechanism for sliding. The viscosity of ice rises with falling temperatures; coupled with a reduced activity index common to low temperatures and light precipitation, this leads to a substantially different mode of flow and a notably different external appearance of subpolar glaciers. The glaciers in Figure 119 obviously differ from those illustrated previously. The small ice field and surrounding individual glaciers behave much as if the ice were a layer of thick molasses poured over the landscape. Figure 120 is another example of such a phenomenon. The small esker along the margin of the glacier at the bottom of Figure 120 suggests that a substantial amount of meltwater must have flowed previously under the glacier. In cold climates, liquid water may be present in channels in ice that is below freezing temperatures for the most part.

97

121. *Subpolar glaciers, Axel Heiberg Island, Northwest Territories*

The glaciers of Figures 121 and 122 are further examples of subpolar glaciers. These all serve as outlet glaciers stemming from extensive ice fields. Both figures illustrate another feature common to these glaciers—the steep, clifflike front of the terminus. There is also a noticeable absence of the prominent lateral and terminal moraines that so often surround the lower ends of temperate glaciers. Both the steep ice fronts and the lack of moraines result from the relatively slow motion and the limited basal sliding. Such glaciers experience a mass balance substantially different from that of temperate glaciers. Precipitation is generally low and is nearly always in the form of snow. Ablation is also low, and meltwater runoff is sometimes almost absent. Where surface melting does occur, it usually is small in quantity and confined to the terminal region of the glacier. The activity index is low. Much of the ablation that does occur is along the periphery, where absorption of solar heat by the dark ground causes local melting and evaporation. This latter feature, combined with the relative immobility of ice, leads to the undermining and breaking off of ice at the glacier edges and consequent cliff formation. The same immobility greatly inhibits the glacier's role in erosion and internal transport of debris; hence the accumulation of morainal detritus is much reduced.

In true *polar glaciers*, subfreezing temperatures are present throughout their mass. Liquid water is restricted to very small amounts of rain or snow melt at the surface. After percolating below the surface, this water freezes immediately. Polar glaciers are frozen to their bed, and the regelation process is completely eliminated. They are found only in the high polar regions, such as parts of Antarctica and northern Greenland.

Although temperate glaciers have received the major emphasis in this book, they actually represent only a tiny portion of the world's glacier ice. Temperate glaciers are the best known and the most frequently visited because in such areas as the European Alps, Norway, and western America, they exist in the midst of populated areas. However, most glacier ice in the world is polar or subpolar. The Greenland and Antarctic ice sheets are the storehouses of some 99 percent of all existing glacier ice, and Antarctica alone contains around 70 to 75 percent of the world's fresh water locked up in ice. These ice sheets are continental in scale, burying whole mountain ranges, basins, and vast plains, and making alpine glaciers seem minor indeed in both areal extent and volume.

122. *Subpolar glacier, "Polaris Glacier," Hall Land, northwest Greenland*

The weight of these ice sheets presses sections of the earth's crust down into the mantle so that the buried land beneath them has a lower elevation relative to sea level than it would if the ice were removed. Similar ice sheets once covered portions of North and South America, Europe, and Asia during the Pleistocene. Other parts of the world, such as North Africa, were ice covered in much earlier glacial periods.

The central plateau of a continental ice sheet, seen in Figure 123, presents the most monotonous scenery on earth. Mountain ranges are completely buried, so that no nunataks break the featureless plains of snow which stretch without perceptible slope or relief for thousands of kilometers. This accumulation of snow, century after century, has formed glacier ice thousands of meters thick. Ice sheets are so large that they make their own climate and affect the behavior of storms over large areas of the earth's surface. At the high altitudes and latitudes necessary to sustain ice sheets in today's climate, the persistent low temperatures (annual mean temperatures as low as $-20°$ to $-50°$ C) keep the air very dry, so that precipitation over Greenland and Antarctica in most places is very small, amounting to only a few centimeters of water equivalent each year. Even with such light accumulation, many cubic kilometers of snow are added to the ice sheets each year simply because of their enormous extent.

The ice in these immense ice sheets slowly flows outward from the central plateaus and descends to sea level, where iceberg discharge eventually delivers it to the sea. Figure 124 shows an edge of the ice sheet in northwest Greenland, where streams of ice spill down from a high plateau into a fjord covered with sea ice. Such land is desolate and inhospitable. Much polar terrain like this is from the climatological standpoint a true desert because of the dearth of precipitation. Despite this, water in large quantities is present as perpetually frozen glacier ice and permafrost.

123. Surface of Greenland Ice Sheet

124. Subpolar glaciers descending from an icecap, Bessels Fjord, northwest Greenland

Figure 125 shows one of the least visited scenic wonders of the world, Petermann Fjord in northwest Greenland. This fjord is far larger in scale than those in Alaska illustrated earlier (for example, Fig. 97). It is 16 kilometers wide, and the vertical cliffs are 800 meters high. This photograph shows the boundary between Petermann outlet glacier, which stretches away into the distance, and sea ice in the foreground. The glacier occupying Petermann Fjord is afloat for a distance of over 100 kilometers, extending far beyond the horizon visible here.

In some places the outlet glaciers of ice sheets discharge large quantities of ice into the sea, where coastal terrain inhibits ablation by iceberg discharge. The glaciers then spread out over the sea surface to form an *ice shelf*. The small shelf in Figure 126 has formed from the outflow of an ice field on the southeastern side of Ellesmere Island in the Canadian arctic. This illustration shows the accumulation areas in the mountains, the ice field, and the ice shelf, all in a single view. Large sections of stagnant "fossil" ice shelves on the northern side of Ellesmere Island occasionally break off and are carried away by ocean currents. Trapped in the Arctic Ocean, they often drift around this basin as *ice islands* for years before they eventually escape to the south to break up and melt away.

125. *Sea ice and Petermann Glacier, Petermann Fjord, northwest Greenland*

126. *Ice field and small ice shelf, Ellesmere Island, Northwest Territories*

A single view of the major ice shelves is not possible in Antarctica, except from a satellite or space ship, for the scale is much too large. Figure 127 shows a section of the Ross Ice Shelf near Kainan Bay. The ice shelf stretches away into the distance, separated from the sea ice in the foreground by the bright line of its terminal cliff, the *ice barrier*. The top of this barrier averages 50 meters above the waterline. The Ross Ice Shelf itself covers over half a million square kilometers of the sea. Large sections occasionally break off to form ice islands which, because of their flat tops, are called *tabular icebergs*. Some have been observed which are 300 kilometers in length.

The glacier features of Antarctica are all on a scale befitting this principal reservoir of glacier ice on the earth. Many of these features have been presented previously for the smaller glaciers and ice fields of North America. The concluding photographs in this book illustrate a few of these features in Antarctica. Figure 128 shows a mountain range from which alpine glaciers coalesce to form valley ice streams. These streams then join to form a broad river of ice, which, in turn, flows into the featureless expanse of the Antarctic Ice Sheet. The sheet itself is also an accumulation zone, in contrast to such a temperate glacier as the Malaspina (Fig. 51), which is predominantly an ablation area with extensive melting.

127. *Floating ice shelf and barrier cliff, Ross Ice Shelf, Antarctica*

128. *Mountain ranges and continental ice sheet, Antarctica*

Figure 129 illustrates antarctic nunataks rising out of the ice sheet. Again the scale is enormous. Many dozens of kilometers separate the nunataks one from the other, and each nunatak itself is a whole mountain range instead of an isolated peak. Figure 130 shows one of the outlet glaciers draining ice from the central antarctic plateau. Vast though this river of ice may appear, by antarctic standards it is but one among hundreds, and a small one at that; a photograph of a large one would show an undulating ice plain stretching to the distant horizon.

129. *Mountain ranges rising above the surface of an ice sheet at Nunataks, Antarctica*

So concludes our short survey of the features of glacier ice. Of the world's wonders, abundant life-sustaining water is without doubt its most unique feature. The oceans cradled life; atmospheric circulation brings moisture to the land upon which many other forms of life developed. As brooks, rivers, and ice, water has modified the landscape into a remarkable diversity unique to this planet. Held in a delicate balance between the boiling and freezing points of water, this world is a habitable spot in a hostile universe. The climate that sustains life as we know it today lies closer to the freezing point of water, and much of modern civilization exists by virtue of a delicate balance between this climate and present snow and ice masses. Were the polar ice caps to melt, much of the world's urban area would be flooded by the rising sea. Return of another ice age would be equally destructive to the temperate zones where modern industrial society is centered.

Scientific study of glaciers, still in its infancy, has already told us much regarding the past geological and climatological history of the world. Present and future growth and shrinkage of glaciers will provide even more useful information, for these changes will be taking place when even the most subtle modifications in climate can be documented as they occur. As the amount of such information increases, the kinds of climates that led to past periods of major glaciation and their subsequent waning can be much better understood. Such understanding is essential if we are to gain a clear concept of the earth-wide environment and man's place in it, for we are—in the geological sense—still in the midst of an ice age that has seen four periods of major glaciation and four periods of glacier retreat, the latter extending to the present. We do not yet know enough about the causes of world climatic changes to be able to tell whether this long sequence of Pleistocene glaciations has come at last to an end, or whether a fifth ice age still awaits us some time in the future. The answers to this question are hidden somewhere in glacier ice.

130. Byrd Glacier, an outlet glacier from the Antarctic ice sheet draining into Barnes Inlet, Antarctica

Glossary of Glacier Terms

The numbers in parentheses give the figure numbers in this book where good examples of the described features can be found. Only the best examples are listed; similar features can often be found in other figures as well.

Ablation. The loss of mass from a glacier, usually as melt and runoff of the meltwater, but also by wind-drifting of snow or the calving of icebergs.

Ablation area. The portion of a glacier situated below the firn line where ice and snow melt exceeds snow accumulation.

Ablation moraine. Unorganized debris on a glacier surface. (47, 89, 91)

Ablation surface. A glacier surface, usually snow or firn, which has experienced appreciable melt, leading to concentration of any dirt present.

Accumulation area. The portion of a glacier situated above the firn line where snow accumulation exceeds melting.

Accumulation layering. The layering in snow, firn, or ice generated by annual cycles of accumulation and ablation (analogous to sedimentary layering in rock). (9–11)

Activity index. The rate of increase of net accumulation with altitude, or the rate of decrease of net ablation with altitude.

Annual mass balance. The annual difference between net accumulation gains and net ablation losses of a glacier. A positive mass balance leads to glacier growth; a negative balance leads eventually to retreat.

Arête. A sharp and sometimes jagged ridge formed by the intersection of two cirques. (108–10)

Bergschrund. Large, semipermanent crevasse at the head of a glacier accumulation zone which separates actively flowing ice from stagnant ice above. (17, 42)

Biscuit-board topography. An upland plateau partially eroded by glacial cirques. (107)

Candle ice. Lake ice that disaggregates on melting into large, columnar crystals. (1)

Chevron crevasses. A series of crevasses along a glacier margin oriented at an acute angle with the margin, usually produced by a combination of compressing flow and side friction. (21)

Cirque. A glacier-carved basin, usually with a steep headwall. (107, 109)

Compressing flow. The flow of glacier ice wherein the velocity seen at the surface decreases downstream. This is the normal state in a glacier ablation zone.

Crevasse. A large crack in the surface of a glacier produced by the stresses of glacier flow. (23, 79–82)

Drumlin. Smoothly rounded mounds, usually of till deposited underneath a glacier and elongated parallel to glacier flow. (115)

Esker. A narrow, winding ridge of till and outwash deposited by a subsurface meltwater stream of a former glacier. (117)

Extending flow. The flow of glacier ice wherein the velocity seen at the surface increases downstream. This is the normal state in a glacier accumulation zone.

Firn. Snow in the state of transition to glacier ice, defined for convenience as snow that has survived one or more seasons of ablation.

Firn edge. The line on a glacier surface dividing firn from glacier ice, or dividing two firn surfaces of different age. (5–7)

Firn line. The average zone of maximum retreat upglacier reached by the snow line each year. The firn line separates the accumulation area from the ablation area.

Fjord. Glacier-eroded valleys that have been inundated by the sea. Fjords may contain existing glaciers. (96, 97, 124, 125)

Foliation. The layered structure of glacier ice generated by flow deformation and recrystallization. (83)

Glacier flour (also *glacier milk*). The extremely fine rock debris that gives a milky color to rivers and a chocolate to green color to lakes fed by glaciers.

Glacier mill (see *moulin*).

Glacier sole. Debris-laden layer of ice at the bottom of a glacier. (28–31)

Glacier table. A rock left perched on an ice pillar by ablation of the pedestal and the surrounding unprotected ice. (83)

Ground moraine. Rock debris deposited underneath a glacier. (35, 115)

Hanging glacier. A glacier, usually small, clinging to a steep mountainside. Also any glacier that terminates abruptly at the top of a cliff. (12, 33, 42)

Horn. A sharp peak of resistant rock left after the surrounding landscape has been eroded away by glaciers. A horn normally marks the intersection of three or more arêtes. Its faces are the headwalls of cirques. (111)

Ice barrier. The front of an ice shelf facing on the sea. Usually a vertical cliff left behind by the calving of tabular icebergs. (127)

Iceberg (see also *ice island*). A piece of free-floating ice broken off from a glacier ending in a lake or the sea. (93, 95)

Icecap. Completely ice-covered area on land, generally in the form of a dome with ice flow radiating outward from the center.

Ice dolines. Glacier surface features caused by drainage of subsurface water-filled cavities and subsequent collapse of the cavity roof.

Icefall. A steep reach of a glacier with a chaotic crevassed surface and rapid rate of flow. (24, 25)

Ice field. An extensive area of interconnected valley glaciers from which the higher peaks rise as nunataks. (112, 113, 119, 126)

Ice island (also *tabular iceberg*). A very large, flat-topped section of an ice shelf which has broken off and drifted out to sea.

Ice sheet. A vast, thick layer of ice which inundates a large area of the earth's surface, i.e., Antarctic Ice Sheet, Greenland Ice Sheet.

Ice shelf. A fixed, floating glacier which extends out from shore over a large area of the sea. (126, 127)

Ice ship. A large pinnacle of clean ice on a debris-covered glacier surface. (89–91)

Ice worm. A small annelid worm which spends its entire life cycle on snow or ice. (84)

Jökulhlaup. A sudden and sometimes catastrophic flow of water discharged from a glacier.

Lateral moraine. A ridge of rock debris along the side of a valley glacier. (33, 35, 45, 46, 104)

Medial moraine. A long strip of debris on the glacier surface, usually parallel to flow, which originates at the juncture of two glaciers. (48–50, 53)

Moraine. A deposit of rock debris shaped by glacial flow and erosion.

Moulin (*glacier mill*). Vertical hole by which a surface meltwater stream enters a glacier. (75)

Nieve penitente. A field of snow or firn pillars produced by an advanced stage of sun cup development. (87, 88)

Nunatak. Land, generally a mountain peak, completely surrounded by glacier ice. (112, 113, 129)

Ogive. An arcuate band or undulation on the surface of a glacier, convex downstream, usually recurring in a periodic pattern. (61–72)

Piedmont glacier. A glacier formed at the foot of mountains by the discharge of ice from one or more confined valley glaciers. (51, 52, 120)

Polar glacier. A glacier that has subfreezing temperatures throughout its mass.

Randkluft. A crack or gap between the edge of a glacier and the adjacent rock wall.

Regelation. Thawing and refreezing of ice through application and release of pressure.

Regelation spicules. Small shards or feathers of ice produced at a glacier sole by the regelation process involved in the sliding of the ice over bedrock. (29)

Regional snowline. The altitude at which annual accumulation balances ablation on a ground surface.

Sea ice (pack ice). A layer of ice derived from salt water frozen on the surface of the sea (not to be confused with *iceberg, ice island, tabular iceberg*). (125-27)

Serac. A block tower or pinnacle of ice or firn formed by the intersection of crevasses. (12, 22, 23, 92)

Snow swamp. A poorly drained area of snow or firn which is saturated with water. (76)

Snowline. The transient lower margin of winter snow accumulation on a glacier. (4, 13, 35)

Soil polygons. Patterns in the ground, sometimes several meters in diameter, produced by subsurface ice in permafrost areas.

Striations (also striae). Shallow scratches and grooves carved in bedrock by glacier flow. (105)

Subpolar glacier. A glacier with subfreezing temperatures in most of its mass. (119, 120, 121, 122)

Sun cups. Cuspate hollows in a snow or firn surface formed by complex ablation processes during sunny weather. (85, 86)

Surge. Periodic, very rapid movement of large quantities of ice in a glacier, alternating with long periods of near stagnation, sometimes resulting in a spectacular advance of the terminus. (37-39, 52-60, 74, 77-78)

Tabular iceberg. An iceberg created when a large section of an ice shelf breaks away. Tabular icebergs have a horizontal dimension very large in comparison to their thickness. They are sometimes called ice islands.

Tarn. A small lake occupying a hollow eroded by glacial erosion. (35, 107)

Temperate glacier. A glacier that is at the pressure melting point temperature of ice in most of its mass. (All glaciers shown in this book are presumed to be temperate except those noted as "Polar" or "Subpolar.")

Terminal moraine. A ridge of debris in front of a glacier terminus marking its maximum stand or position of readvance. (16, 35, 118)

Terminus. The lower end, or snout, of a glacier. (21, 32, 35, 38, 43-44)

Tidewater glacier. A glacier that terminates in the sea. (94-96, 124-27, 130)

Till. Unsorted rock debris deposited by a glacier. (116)

Trimline. The limit of erosion or modification of vegetation by a glacier at a position of former advance. (34, 35)

Unconformity. A discontinuity in accumulation layering. (11)

PHOTOGRAPHIC CREDITS

Photocomposition in Medallion
by York Graphic Services, Inc., York, Pennsylvania.

Duotone lithography by Graphic Arts Center, Portland, Oregon,
on Warren's 100 lb. Cameo Brilliant Dull paper.

Binding by Lincoln and Allen Company, Portland, Oregon;
cloth by Joanna Western Mills Company.

Designed by Audrey Meyer.